The Forgotten

Submarine Pioneers

Richard M. Jones

ISBN 9798326813688

Copyright © 2024 by Richard M. Jones

The moral right of the author has been asserted.

All rights reserved.
No part of this publication may be reproduced, distributed, or transmitted in any form or by any means, including photocopying, recording, or other electronic or mechanical methods, without the prior written permission of the publisher, except in the case of brief quotations embodied in critical reviews. For permission requests, please email the publisher.

A CIP catalogue record for this book is available from the British Library.

Richard M. Jones
Website: https://shipwreckdata.wixsite.com/richard-m-jones
Blog: www.disasters-shipwrecks.blogspot.com
Email: shipwreckdata@yahoo.co.uk | **Facebook**: Richard M Jones
Instagram: @wreckmasterjay | **YouTube**: @RichardMJones-js7fz

Front cover photograph by Aarón González
unsplash.com/@aarez | https://aarez.es
Instagram @aarez.design and @aarongonzvlez

Juliette Jones
Editorial Services

Edited by Juliette Jones Editorial Services
juliette.jones@outlook.com | www.juliettejones.co.uk
Facebook/Instagram @JulietteJonesEditorialServices

This book is dedicated to James Ellis Howard and Charles Seymour Allan

Also by Richard M. Jones

The Great Gale of 1871

Lockington: Crash at the Crossing

Capsized in the Solent: The *SRN6-012* Disaster

End of the Line: The Moorgate Disaster

Collision in the Night: The Sinking of *HMS Duchess*

Royal Victoria Rooms: The Rise and Fall of a Bridlington Landmark

RMS Titanic: The Bridlington Connections

The 50 Greatest Shipwrecks

Britain's Lost Tragedies Uncovered

The Burton Agnes Disaster

When Tankers Collide: The *Pacific Glory* Disaster

The Diary of a Royal Marine: The Life and Times of George Cutcher

The Farsley Murders

Living the Dream, Serving the Queen

Around the World in Shipwreck Adventures

Boleyn Gold (Fiction)

Austen Secret (Fiction)

Gunpowder Wreck (Fiction)

Cretil the Cat (Children's book)

Lost at Sea in Mysterious Circumstances

A-Z of Bridlington

Shipwrecks of the Solent

Contents

Introduction 1

The Development of the Submarine 5

Charles Seymour Allan 15

James Ellis Howard 32

Epilogue 45

Acknowledgements

Sources

Author Biography

Introduction

In 2022 I began the research for my final year of a Master's degree in Naval History and was looking for a new and interesting project to write about for my dissertation. I needed something that had never before been studied and also a subject that would ignite both questions and conversation with other academics, as well as become a record for future historians to use. The whole story of how the submarine developed over many hundreds of years was already a fascinating subject, who would have thought that the first ever serious testing of an underwater craft would date as far back as the 1600s and have occurred right in the middle of London? Not only that, there were issues as to the reliability of some sources when it came to famous quotes and incidents; most notably was a widely accepted quote from Admiral Wilson, in which he supposedly called submarines "damned un-English," but further research suggested that he may not have ever said that in official circles at all. It is possible that this could have been nothing more than a throwaway comment at a cocktail party or joking amongst his peers. The fact that the phrase stuck and has since been repeated in numerous books confirms that if you tell something often enough it is then published as truth, whether it is or not.

But the more I read about the engineering pioneers who built and tested these remarkable contraptions the more I wanted to dig deeper. As somebody who has researched shipwrecks since I was a boy, I was already familiar with so many of the stories, that of the Confederate *H.L. Hunley*, the sinking of the liner *Lusitania* by a German U-boat, the loss of the Russian nuclear submarine *Kursk*; there were so many tragic events but at the same time so many reasons for wanting to learn more – the exploration of our vast underwater world being a reason I studied the seas and oceans so hard as a kid, wanting to do the same. My all-time hero was Dr. Robert Ballard, who found a host of incredible things on the seabed, will always be remembered for the discovery of a number of historically significant shipwrecks including that of the *Titanic* and the *Bismarck*.

As with all of the books I have written, I have tried to find something that is obscure and needs more work and this was the same attitude I would carry over into my dissertation. While searching for any lost and unpublished information on the early submarines I first located some old newspaper articles from the late 1890s about two submarine designers that I had never heard of; that soon got me searching for more information on them. I found that I could not find a single book to have ever mentioned the names of James Ellis Howard or Charles Seymour Allan; besides these news reports, there was nothing. I came up empty in the archives too.

What happened afterwards was a painstaking search around the world for any information on these two people, both of whom seemed to be based in Australia. Thankfully I did manage to find little bits, here and there, which when put together built up a fairly good picture of two innovative pioneers who threw everything they had into their inventions to try and make them not only work successfully, but then they would have to change quickly from inventor to marketer as they then had to sell their own ideas to the British military. Both men embarked on a long journey from their homes Down Under to the bustling streets of London where they would catch the eye of the Admiralty and each finally get a chance to show off their creation and make their fortune.

It was fascinating to learn about this period of history and exciting to think that nobody had written about these two people before. I always said that once I had finished my degree I would write further about Howard and Allan in a small book and delve deeper into their lives in order that their achievements are finally recognised and made public.

In October 2023 my results came through; I had thankfully passed my Master's degree. This is the long awaited book on the life and times of The Forgotten Submarine Pioneers.

The Development of the Submarine

The dream of man under the sea has been a subject of conversation since at least the years around the reign of Alexander the Great, when he was apparently lowered into the sea in a glass barrel in 333 BCE, although whether this actually happened has long been open to debate. But the fact of the matter is that the idea of a submersible craft had been in the pipeline for many hundreds, if not thousands, of years. The first ever testing of a submarine designed unit was by Dutch inventor Cornelis Drebbel between 1620 and 1624 while he was working for the English Navy (it would be several more decades before the title Royal Navy was given). He designed and tested a contraption that could travel down the river Thames and submerge to several metres, apparently even inviting King James I on board to try it out. Whether any of this is what really happened remains to be debated, but the fact is that Drebbel had achieved what many had only dreamed of, although this was not built upon and it would be many more years before further submarine designs would be seriously considered to be worth the cost. But the dream was alive and inventors got to work in trying to design their own underwater boat and somehow install some kind of enthusiasm amongst the general public and militaries worldwide.

By the time of Jules Verne writing *Twenty Thousand Leagues Under the Sea* in 1870 there were already periods of history that were showing just how useful a submarine could be when it takes the fight against an enemy ship to a whole new level; at this point it was all about the ability to conduct warfare and very few people were talking about the undersea exploration benefits to it.

David Bushnell's *Turtle* was the first recorded attack on a warship, this took place on 6 September 1776 when the American War of Independence was at the height of hostilities. The craft itself resembled two turtle shells put together (hence the name), only just enough room inside for one very uncomfortable human being who would pilot it across the water using hand cranked propellers. Setting off across the harbour of New York, pilot Ezra Lee single handedly brought the *Turtle* to the anchorage where the British fleet lay and then slid under the hull of Admiral

Howe's flagship, HMS *Eagle*. Screwing an explosive charge to the underside of the warship, the *Turtle* would then back off and head back to base, but the attempt failed to secure the charge to the bottom of the ship as the sub retreated and so the *Eagle* remained afloat to tell the tale. Doubt has been cast as to whether this did indeed happen as the laws of buoyancy would have kept the *Turtle* afloat and even if it could submerge under a ship, the act alone of screwing the explosives upwards would have pushed the sub downwards. Not only that, at no point that night did anybody from the *Eagle* make a report of a craft nearby, although some reports said that a craft was chased away from the nearby area by boats. True or not, the story of the *Turtle* went down in history as the first submarine attack, but it would be almost a century before an actual attack on a surface vessel was successful.

Fast forward to 1864 and the United States of America was once again at war but this time with itself, as the civil war saw the two sides, the Unions and Confederates, battling it out and having some major clashes that are still studied by historians even today. But in the background were inventors like Horace Lawson Hunley who built his own submarine in 1863. Designed to carry a crew of eight, the cigar shaped *H.L.*

Hunley would be propelled by pure muscle power as they manually operated the long handle that stretched the length of the inside of the submarine and in turn rotated the single propeller until they were within sight of their enemy. Unfortunately, during several trials the *Hunley* flooded and this led to it sinking completely, killing five crew on 19 August 1863; after it was recovered it went on another similar trial on 15 October, in the same year, where once again it sank, this time killing all eight on board. Once again it was salvaged and put back to sea, this time on a real mission to attack the surface warship USS *Housatonic* off Charleston, South Carolina, whereby she rammed the ship's side and backed away, leaving her torpedo jammed in the hull. The resulting explosion ripped the *Housatonic* apart and she quickly sank, but the fate of the *Hunley* was unknown from that moment on. She had vanished and was presumed lost in the blast wave; the wreck was eventually located in 1970 by explorer E. Lee Spence, which finally confirmed she had been sunk that day in the explosion not far from the blast. *Hunley* was salvaged in 2000 and is today in a museum, her eight crew given a burial upon recovery, over 130 years after they had been lost.

But *Hunley* had now officially become the first submarine ever to sink a warship and that fact alone propelled this small craft into the history books. Incredibly, this feat would not be repeated until the start of the First World War, almost five decades later. In the meantime a number of people wanted to design, invent, test and produce a working attack submarine for the military (and it is shown throughout this period of history it did not matter which country's military it was sold it to either) and so these designers worked hard to be recognised for their innovative ideas and dogged determination.

While the American inventors were busy making submarines during their wars, there has been very little evidence that any other country was even contemplating the idea. That is until author Jules Verne changed all this, the star of his book being Captain Nemo aboard his futuristic underwater vessel *Nautilus*. As a concept this was an idea well ahead of its time and the public loved it. The name of Verne's submarine was not the first *Nautilus*, Robert Fulton had actually built a craft with that name in 1800, long before any fictional version and well before the idea of a genuine attack submarine had even begun to be taken seriously. Again it was the anti-British feeling that led to this American inventor wanting to use his idea against the Royal Navy, so much so that he had moved to France in order to work with Napoleon who had begun his rise to power in 1799. A lack of progress in the development stage forced him to flee to Britain where he tried selling his design under an assumed name, of course he wouldn't want the British to know who he had been working against all this time. Ironically it was Prime Minister William Pitt who saw what he was doing and took a keen interest in what he had to say; Pitt instructed the Admiralty to investigate his work and Fulton gave a demonstration whereby he blew up a brig named *Dorothea* using a submerged mine – the result being that Pitt immediately wanted to buy the submarine. Unfortunately for Fulton the Admiralty rejected it outright, despite the Prime Minister's enthusiasm. Had they pursued this project, the Royal Navy could have been the developer and user of the first real submarine a full hundred years before anybody else.

Meanwhile, over in Manchester, the Reverend George Garrett was designing and building his own version of the submarine, although this had a very different look about it than any other previous design. A 27-year-old Clergyman with a vision, he had built from scratch a remarkable looking craft consisting of a cylinder with sharp points extending out each end, a conning tower rising up the centre section, one single propeller and the most amazing part of it was the fact this particular submarine was fitted with a steam engine. He named it *Resurgam*, Latin for 'I will rise again'. Getting to this stage of his project had been a long struggle with different tests being carried out, but even when he had to cut short his previous trials, due to the extreme heat of the coal fired engine, the submarine was put to sea and still performed underway independently for an hour. Knowing that there were issues with the build-up of toxic gases within the compartment, Garrett went back to the drawing board and developed a breathing device to compensate for this. Unfortunately, when under tow for further tests on 25 February 1880, the submarine was overwhelmed in rough seas and sank off the coast of Wales. This submarine was the closest the Royal Navy had come to showing an interest in adopting any kind of submarine, but after the sinking of the *Resurgam*, they quickly lost interest once again. Garrett, however, was not one to back down from an idea and soon had contracts in Sweden, where he would work closely with his Scandinavian counterpart Thorsten Nordenfelt, to make yet another successful project; in this case it would be for the Nordenfelt Submarine Boat Co and history would see that Nordenfelt took all the credit for these inventions before the Swede went on to set up an ammunition and machine gun outfit. Garrett found himself once again on the move, travelling this time to go and work in Turkey where their military was so impressed by his work that they made him a naval commander. A second version of his submarine was built in Barrow but during the delivery voyage to Russia, once again, his boat sank en route. Things took a turn for the worse over time and he ended up dying in poverty in New York at the age of just 41, the realisation that he had invented the world's first mechanically propelled submarine only being really known in detail when the wreck of the *Resurgam* was confirmed as being found by divers off the coast of Wales in 1995.

Then along came John Holland. He was just 23 when news filtered around the world that the USS *Housatonic* had been sunk by a submarine and at this time of his life Holland was nothing more than an Irish monk living in a monastery with a quiet life ahead of him. Finding the notion of an underwater attack craft a huge fascination, he became increasingly distracted from his work within the brotherhood and in 1873 he left the order to pursue a new life on the other side of the Atlantic, ending up as a teacher in New Jersey. A far cry from living in a religious compound, Holland would now make his dreams happen and in doing so change the course of maritime history. Although he was now in the United States, strange as it sounds, this is where the birth of the very first Royal Navy submarine began, at least in its design stage. He submitted a very basic underwater craft design to the American Government, but much to his dismay they rejected it straight away. Despite the setback, unbeknown to him there were a few other people following his work, these people were known as The Fenian Brotherhood. This organisation was hell bent on fighting the British for an independent Ireland and they were looking closely at Holland's submarine design with a keen interest. His background did not go un-noticed and he was, after all, one of them and they had no problems with finding enough funds to allow him to develop his new weapon of war for their own use. Over the next few years he designed and built several prototype submarines, but frustratingly for him, the Brotherhood were not as loyal to each other as he had first thought. Arguments raged within the ranks and a group of them stole two of his prototypes, with Holland being forced to scuttle a third otherwise they would have had all three. One of the two stolen boats then sank anyway and the only other one could not actually be operated without Holland himself being there – the thieves had literally not a clue how to work the submarine and had to give it up as a bad job. This entire submarine project for them faded away and Holland could now move on from this episode very quickly and try to find funding from elsewhere. But while Holland was busy with his projects, two men in Australia had already started to make waves with their own designs; this is where Charles Seymour Allan and James Ellis Howard come into the story. The two of them could very easily now alter the path of the submarine design.

Charles Seymour Allan

By the start of the Twentieth Century the public were becoming accustomed to hearing about the new underwater inventions in the papers. The *Illustrated London News* in 1901 described 'much interest in the submarine boats ordered by the British Government'. While the article makes no mention of who exactly has this interest it is easy to assume that it is just referring to the general public as well as those who work for the Navy both on the ships themselves and of course those who work within the Admiralty buildings; although the size of the article does not convey the excitement that it may have done, perhaps this is just the sign of the times, for if it had occurred in the modern day it would most likely be a double page spread on the design taking up the middle pages. The same newspaper five years earlier had featured an image of what one of these submarines would look like as it gave details about a model exhibited at the St George Swimming Bath in London and what it described as 'a submarine torpedo-boat and blockade runner' hosted by a man named

Mr Seymour Allan (often mis-spelled as Allen), who was catching the attention of 'the Naval Attaché's of most of the foreign embassies.'

When Allan had gathered all interested parties together, he then went to grab their attention so that he could now introduce his new invention to his eagerly awaiting audience. What followed was a demonstration of a model of his new design of submarine and a lecture on the gyroscope that would be helping to navigate and steer the boat whilst underwater. Lowered into a swimming pool, the model started off by being set off in full view of the anticipated onlookers, the craft being seen to submerge in the pool as planned and then being put through a number of tests whereby the prototype submarine was sunk, raised, allowed to hover and then remain stationary upon demand. He turned to his audience and announced that a full sized working version of the model that he was to build would be around 80 feet long and displace 127 tons; it was clear that Allan had done his homework when it came to the vital statistics of his design. While this event was a success the newspaper was already convinced that this was the future of underwater warfare and they already spoke highly on the vast potential that Allan offered for the Royal Navy –

'Altogether it was a most successful demonstration of a principle which may be expected before long to be embodied in the form of an actual service vessel equal in size to one of our first-class torpedo-boats.'

The man who was doing this demonstration was Charles Seymour Allan, a man who found himself now a part of the submarine development story for the Royal Navy but over time has never been mentioned in the historiography of the submarine service nor featured in any of the museum exhibitions or history books.

He started his life in Britain, the son of John Brackenbridge Allan of Eddlewood in Scotland and at some point he made a journey across the oceans, heading south to Australia where he would go on to get married at St Stephen's Cathedral in Brisbane, on 21 November 1894, to Mary Eveleen Perkins, eldest daughter of the Honourable Patrick Perkins of Brisbane; according to the papers a very large crowd attended, the bride being given away by her father and accompanied by five bridesmaids. The celebration party afterwards was held at Allan's new in-laws where 150 guests cheered them on, the newlyweds soon leaving by mail train for Sydney that same day. The *Cobram Courier* talks about him in their columns as being Australian and at some point his surname was misspelled and in other cases made double-barrelled into Seymour-Allen by the press for some unknown reason. What is interesting to note here is that it was a whole year before the submarine tests in London that the *Courier* had described his invention as such –

'The Brennan torpedo, of course, we know, but not the Seymour-Allen Submarine boat, which quite recently has been tried as a model at Melbourne before the Governor, the Naval Commandant and a large party of naval and military officers.'

So he was already making waves, so to speak, with his ideas and throwing them around to whoever would listen, rightly so when he had something so revolutionary to shout about. An illustration drawn for the Illustrated London News in 1896, but with very little other information to go on, shows Allan as a man with a large moustache, balding on the top of his head, wearing a tuxedo (or similar) explaining a large submarine model that did resemble loosely the *Nautilus* from the Jules Verne book. Why the papers didn't report further on this, considering the amount of space the images took up in the newspaper, is anybody's guess but at least it gives us the only image for us to know what Allan looked like – and they spelled his name correctly!

It is around this time that Australia seems to be the homeland of a lot of the submarine inventors according to these contemporary press reports, but very little is mentioned in the British newspapers other than the swimming pool test the year later. The Australian news reports go on to describe the speed at which the test model can go as it propelled just as fast underwater as it did on the surface, *The Evening News* (Sydney, New South Wales) does make a report of further tests in front of a Captain Castle RN as well as a host of other dignitaries such as members of the Marine Board and a number of military and naval officers. In this test Allan not only demonstrated the model working, but started to reveal the price of building the real thing. He turned to them and asked them to consider the construction costs of an ironclad warship which he stated would be in the region of around £1,000,000 compared to how much it would be to build his submarine at around £12,000 to £15,000 and then the annual running costs on top of that. Considering the cheaper one would destroy the more expensive one, in his opinion it was common sense to invest in his submarine, but then again he was giving them a sales pitch so he would obviously be highlighting how his craft was better, completely missing out any details regarding the costs involved in the amount of trials and tests that would have to be undergone in the meantime, not forgetting the crew costs, weapons once it was at sea, general maintenance and repairs and that is just for starters. He asked them to consider the destruction of 'the most costly ironclad ever constructed,' before showing them his model at work. Whatever he said behind the scenes he was obviously passionate about his invention and he seemed to be bringing people around to his way of thinking as time went on.

But it was not only model designs he was willing to prove, another report in the *Daily News* dated 15 November 1894 suggesting that Allan himself would build his own submarine from scratch and he was so confident that he was more than willing to crew the finished submarine himself to prove that it worked as planned, a man not afraid to put his money where his mouth is. He said that there would be no plans to be taking his invention abroad, instead a full size working submarine would be built right there in Australia, where he would proceed to test out his theories and pilot it out to sea and then the plan was to sink a hulk by the use of a torpedo. All this would be done without telling anybody he was going to do it first, clearly giving an element of surprise and for the shock factor to set in when people had seen the results. At this point he figured that people would start coming to him instead of him travelling around trying to sell the idea. This was clearly a good concept as long as the building and testing of this actually went ahead and was successful, but Allan's submarine design and testing then suddenly vanishes from the newspaper reports in favour of the work that John Holland was carrying out, with no further record of his proposed sinking of the hulk ever taking place.

After all that effort and expense, his design never makes it into any military as far as the records go. There are no more reports of him showing off any of his models and in more than one report he is listed as being named simply as 'Seymour Allen' making it seem like that is his full name. There is, however, one incident that was related in Australia that fits in with his life and time line, but it is nothing to do with submarines. It was in the June of 1895, less than a year after he married Mary, that a libel suit in Sydney against a Mr Daniel Lehane Willis commenced, whereby Allan was suing for £10,000 in damages brought upon by an article that Willis had written in the magazine *Truth* which was titled 'Duped and Deserted. A Brisbane Heiress married an English Swell, who disappeared with her jewels, and leaves no substantial truth behind'. The story goes that Allan was an Englishman who had been married to a rich heiress in Brisbane, but that he had also been married to another woman back in England in which he had left her with six children when his bigamy was found out, returning to Australia to elude justice, taking with him a number of jewels. As none of this could be proven and the fact that Mary accompanied him on his trips, the jury awarded Allan damages of £1000, a far cry from what he was hoping for; but still, he won the case. There is no mention of him here being an inventor but this was the same man that had only recently wowed the crowds with his submarine model and now found himself making headlines for all the wrong reasons through something that wasn't even true. Back in those days it was no small feat to travel across the world, so having two people with the exact same name in the same time period may be just nothing more than a coincidence, or it could in fact be that Charles Seymour Allan was just a man who had a very eventful life and was caught up in a case of mistaken identity.

In 1895 Allan was named in a divorce case back in Britain when a man named Templer Edward Edeveain had found that his wife Annie Elizabeth had committed adultery with not only Charles Seymour Allan but also at some point with another man named Captain Harry Holt. Annie had met up with Charles and in 1891-1892 they had arranged several rendezvous in hotels around Britain, certain places named in the documents were the St Pancras Hotel in London, the Star and Garter Hotel in Richmond and the Royal Hotel in Leicester. The divorce was finalised on 4 November 1895 after over two years of legal wrangling, the petition having been filed on 3 August 1893, presumably once Templar had found out about her affairs. As there is nothing else available for us to see regarding this we can only speculate on the circumstances surrounding this part of Allan's life. This all happened long before his own wedding and by the time it was all over he was married to Mary and had been for a year, one only wonders what she may have thought about all this, whether Allan had knowingly had a love affair with a married woman and if he did then why? In a time when women's voices were rarely heard, particularly against their husbands, we may never know what really went on between Annie and Charles, or more to the point Annie and her husband.

At some time or another he moved into 1 Lovelace Gardens in the London borough of Surbiton and he is listed in the 1901 census as living with his wife, listed in this case as Eveleen, his two sons Patrick and Jerome and as well as a servant.

One thing is for sure, he was accustomed to travelling to Bognor Regis on the south coast to visit his family, accompanied by Mary, sometime during the period 1895-1901, so the bigamy story that was revealed by the *Truth* article cannot be true because Mary was with him on his travels the whole time. Charles's mother died in October 1900 and after that he never returned to Australia after this visit; he died in the Sussex County Hospital in Brighton on 30 September 1901 at the age of just 44 after a short illness, his death certificate listed urethral abscess, extravasation of urine and heart failure as the cause of death. His sister from Weybridge registered his death two days later and Allan's address is listed as 54 Western Road, Hove.

Charles's beloved wife Mary returned to her home country with their two sons Patrick (aged 3) and Jerome (aged 2), both of whom had been born in Britain during their time spent here with their parents. Patrick would suffer his own tragic demise when he was lost in a mustard gas attack during the First World War and Mary died on 21 October 1942 and was laid to rest in Waverley cemetery, Australia.

At the time of writing it is unknown where the last resting place is of Charles Seymour Allan, after contacting a number of places regarding death registers it is not listed as anywhere in the Brighton and Hove area and indeed may not even be in Britain.

A 1604 portrait of Cornelis Drebbel who tested the world's first submarine on the river Thames.

The Illustrated London News shows Charles Seymour Allan demonstrating his new submarine model to an eager crowd in London in 1901 (Mary Evans Picture Library)

Hobart, Tasmania, 1900 (photo by John Watt Beattie)

James Ellis Howard with Rosetta Eliza (centre left) and their daughters Ethel May (top) Florence (bottom). (Daphne Purdon photo)

James Ellis Howard (Daphne Purdon photo)

Howard shares certificate for his submarine company (Daphne Purdon photo)

The grave of James Ellis Howard (to the left of Heard grave), Lambeth Cemetery, during a visit by the author in 2024.

John Philip Holland became the first person to supply the Royal Navy with submarines after his success with the Americans. This photo shows him stood in the hatch of the Holland I.

After sinking in 1913, the wreck of Holland I was raised in 1982 and has since been preserved and put on display at the Royal Navy Submarine Museum in Gosport, Hampshire (author photo).

Submarine design has come a long way in just over 120 years. This is the future submarine AUKUS project announced in 2021(BAE Systems).

James Ellis Howard

While the life's work of Charles Seymour Allan was fading, the other side of the world once again had a surprising addition to the submarine technology race when the Welsh newspaper *Rhyl Record and Advertiser* announced that 'the Admiralty had been offered a new submarine boat ... Its claim to be superior to other inventions of the kind'. This story was followed up a few months later by the *Daily Express* announcement in September 1900 that the man involved in this new boat, Mr. James Ellis Howard, had developed a 'Terrible Destroyer' in his hometown of Hobart, Tasmania. After overcoming a number of difficulties, he had now made the claim that he was ready to trial his submarine and organise an attack on a surface ship to prove that the accuracy and skill of his craft was worthwhile.

Shaped like a cigar and having the parts constructed in absolute secrecy in a number of different workshops, Howard had already given enough details to the reporter about his methods of launching a torpedo attack and the tactics that would be used, namely the torpedo being attached to the ship's bottom using a suction device and then afterwards the submarine retreating back. This seemed very much like what happened with the *H.L. Hunley* in both design and methods, given also that the report stated:

'All this sounds terribly deadly. If Mr. Howard can do one half of this, the death knell of navies, so far as large battleships are concerned, is within reach.'

Just reading this quote alone gives an idea of how confident some of the observers were that this was indeed a game-changing design, but with it still being in the early stages it does seem a little over-enthusiastic considering no real results had yet been proven other than what is written on paper. But it is the excitement and drama of the headlines that would sell the papers and it attracted enough attention to get people to read these reports, so it did not matter if it worked or not, as long as people read about it and then subsequently discussed it into the bargain. Howard had been working on boats for a number of years, a news article in 1874 spoke of a barge that had been built by him that was thirty six feet long and would be placed on what it described as 'the Huon trade' which is a region of Tasmania on the eastern coast of Africa. Twenty years later the French, German and Japanese navies were all in contact with him at some point or another and were offering him large sums of money for his innovative ideas and were 'cultivating an acquaintance' for the skills of a man whose father himself had served in the Royal Navy and the Kent Coastguard.

However, a strange incident in 1899 was mentioned in the *Sydney Evening News* when both James Howard and his model prototype submarine suddenly went missing from the local area, it then being reported that Howard was later arrested in Melbourne over the affair, his ten year old daughter being with him at the time. What exactly Howard, along with his co-partner, were being arrested for it only says that it was a simple case of larceny, suggesting that it was the submarine itself that they had taken. Apparently with it having a value of £293 this had made the disappearance of the craft all the more reportable to the police, but this adds that little bit more mystery as to his involvement in this submarine design and in another case it perhaps throws a further level of secrecy upon it. Not only that, but these reports make no mention of what had caused him to take such drastic action in the first place, remember almost the same thing had happened with Holland and the Fenians when they stole Holland's work. But the work done by Howard was at last being noticed, the *Brisbane Telegraph* reporting on 30 June 1900 that:

'Its claim to be superior to other inventions of the kind is based on the fact that when the torpedo is discharged it fastens itself by means of a suction arrangement against a ship's bottom and has a time fuse, thus giving the submarine boat an opportunity to get away ... Mr Howard is now in London preparing to put his invention to the test, if so desired by the Admiralty.'

Several months later the British newspaper *Daily Express* once again took up the story, reporting on 7 September 1900 the dire need for the Royal Navy to possess a submarine:

'The French submarines are better than those of Russia, Italy or the United States, but even with this superiority our near neighbours have not solved the problem of navigation and delivery, as Mr. Howard's claims to have done.'

This is all well and good as a comparison between navies, but this still remains just a single person's opinion being published in a newspaper article, one can only imagine how that same reporter would change his tune if he was, for example, an American. Any reader of this at the time would have been impressed by the story of the submarine development, but looking closer this reporter claims that the problems of navigation and delivery have been solved, but does not say what these problems are or what was done to solve them, nor has Howard claimed any of this to our knowledge. This could quite easily be a general quote from the designer taken out of context to a reporter who has no technical knowledge of what is being said or know of the history of such equipment.

But just who was James Ellis Howard? Born in 1836 in Dover, Kent, he had grown up with the sea in his life one way or another, what with his father's Coastguard and Navy connections, but soon found himself emigrating away from his home country. At some point he made Australia his permanent home and was soon settling down with his wife and, later on, his children. His personal life is somewhat hazy at this point, having seemed to have no problem fathering children and not all of them being to his wife either!

At 32 years old, on 16 June 1868, Howard married a 20 year old Australian girl named Margaret Elizabeth Brereton in Hobart, Tasmania, and they had three children together. At some point he then went on to father a further five children with a woman named Rosetta Eliza Hurst. His signing of his Last Will and Testament on 19 October 1898 has named all his children plus Rosetta and the fact that he was living in New South Wales. His personal life does seem very complicated, more so when he appeared in court in 1876 for neglecting to pay maintenance to his children and again a year later for the same issue. It is interesting to note that in the register of births when his daughter Susan was born in 1890 he is listed under occupation as simply 'carpenter'. Not only that, Rosetta seems to have taken the name 'Howard' but at this point yet there is no evidence either way to say that he was or wasn't still married to Margaret. Rosetta had five children to Howard. Ethel, Florence, Daisy, Susan and James, although tragically Daisy had died at the age of just under four months old in 1889. There is a family photograph of Howard posing with Rosetta and two of his children, Florence and Ethel, showing a man who has a thin and shaven face yet sitting tall and proud. As Florence was born in 1887 and she is looking to be around ten years old, this image most likely shows Howard when he was in his early-60s after having an already full life and with seemingly still very much more to do. He obviously had a lot of feelings for Rosetta posing in a family photo like this as well as naming her as the benefactor of his will instead of his wife.

Aside from his love life, Howard seems to be always hard at work. On 28 May 1900 a company was set up named the 'Howard Submarine Boat and Torpedo Inventions Company, No Liability' in Melbourne, by now he was at the age of 64, and this company was registered as being managed by a Mr. Alf C. Horsley. By November of that year Howard was making headlines yet again, this time in the Melbourne newspaper *The Argus* when they reported that a company had been formed in the early part of the year so that he could be sent to London to negotiate the sale of his invention to the British Admiralty. The report spoke about the fact that there had been a successful trial of his submarine at Sheerness on 19 November 1900. Further investigation shows no records of this trial, so whatever happened it didn't seem to make the waves he was expecting, and very little can be found giving any details other than these flimsy press reports. There was talk of The Holland Company approaching Howard to discuss his invention and perhaps amalgamation, Howard himself proposing to float a company if the Admiralty reject his boat for trials – with half the capital needed coming from his homeland of Australia. Not only that, he planned for two types of boat to be built – one for the harbour and one for the deep sea, but size now being seen as a factor in these comparisons. The way the *Daily Express* writes about these articles (there seems to be only this paper in Britain that really takes on this story, the rest just have a small paragraph here and there), it seems that there was excitement in the air at the prospect of Mr Howard and his submarine finally becoming a reality.

Examining another Australian newspaper gives us a little more detail about the Howard Submarine Boat and Torpedo

Inventing Company itself, which by now had the registered address as 480 Bourke Street in Melbourne, claiming that Howard had tested his submarine successfully in the presence of an Admiral Pearson who 'stated that it was a wonderful and most important invention,' before expressing satisfaction at the outcome of the trial. Where all these tests were carried out and just how many people were present, other than the Admiral, is something to be only guessed with the lack of records. Of course the papers never cited their sources for every quote and snippet so it leaves us to wonder a century on just how much was simply hearsay, truth or even completely made up. But whichever way you look at it Howard was making progress.

But then everything seems to go quiet once more. The *Evening Telegraph of Australia* publishes a report on 8 May 1902 calling Howard the inventor of 'A submarine destroyer' and that he was heading back to England after returning home in the spring of 1901 after he had failed to get a satisfactory hearing from the Admiralty. This then suggests that his new invention is not so much a submarine in itself but actually a surface vessel that is instead designed to destroy the submarines – a very early innovation in anti-submarine warfare. Whether this is just the terminology being lost in translation or not means that the modern historian has to take these few reports at face value unless other evidence can show otherwise. The problem Howard seems to have had is that any official tests have to first make public details about his invention which he had previously declined to do. For him the secrecy of the boat was paramount, especially with so many rivals coming out of the woodwork. This was not a military project, it was a private individual that was selling the idea and design to the military, so secrecy for the Royal Navy was nothing to do with them. It was Howard himself that would have to keep it all under wraps so that others didn't steal his ideas. At this point in his life he is now seen to be even using his own money to build and test the boat, while keeping in touch with a Captain Bacon of the Royal Navy, who was an Admiralty submarine expert based up at Barrow-in-Furness. With rumours of Japanese interest in his invention, he claimed to have solved all previous issues of how the submarine craft operated underwater and deployed the so-called torpedo. However, things don't seem to have gone his way as the news of his work dried up very quickly within the press reports and his invention is never spoken of again, almost like he was never there in the first place.

What happened to the invention that was making the headlines and why did it go from the hot topic of Tasmania to suddenly having no real evidence of its existence is now down to what records are available to corroborate his life's work. James Ellis Howard is one more piece of submarine history that only made it so far and then vanished, not even having a note in any of the history books. What is clear here is that only the contemporary press reports now exist that even show Howard had ever approached the Admiralty, let alone actually tested any submarines, the archives have no reports and no testing logs for the historian to actually see what went on during this period and to make matters more complicated there is never an actual name given to his submarine invention that can be linked back to him. But it was reported for several years that he did actually test his submarine, at the least the Sheerness report confirms this. But where did this submarine come from, more to the point, what happened to it after the demonstration?

Howard's life suddenly drops out of the limelight, there were very few reports of his whereabouts over the next two years until he is then listed in another Australian newspaper as dying at the age of 67 in London on 11 May 1903. His death was listed as being a number of factors including double aortic, vascular disease, angina pectoris (chest pains) and cardiac failure where he died at his home at 48 Glenelg Road, Brixton in the presence of his family; his 17 year old daughter Ethel registered his death the day after. His funeral was held soon after, with his body being taken to his final resting place at Lambeth Cemetery in the Tooting area of London.

Very little is known of the women who were the mothers to his children. Howard's lover Eliza Hurst died on 12 January 1926 in Hobart, Tasmania at the age of 62. His wife Margaret Howard lived until the age of 82 when she died in November 1929 in Victoria, Australia. They have both vanished into obscurity, their link with James Ellis Howard now just a footnote in their lives, studied by only the relatives of the Howard family who want to know their family history and trace back any ancestors.

For the man who had invented and tested what could have been the Royal Navy's first submarine, his life is almost forgotten, as is his work. In February 2024 I paid a visit to the grave of James Ellis Howard and after a bit of searching it is clear that today he lays in an unmarked grave, numbered 356 E2. Just a small stretch of grass covering the body of a man who could have been so famous in history; he came so close but not close enough.

Epilogue

Today the work of Charles Seymour Allan and James Ellis Howard is not recognised anywhere in the world. It was only searching through old newspaper articles looking for things on John Holland that I found it in the first place and I am glad I did. These two men were great inventors, the entrepreneurs of their day with good ideas and a drive for change. The fact that they were then overshadowed by other people's work is no slight on their efforts, it is just sad that their lives have never before been noticed or their story told before. What could they have done further to make bigger headlines? We can only imagine today what would have become of them if a rich backer had funded their projects and allowed them to continue their work. I am just pleased that I found the few things I did and that the information I have managed to gather together is enough to show the world who these people were and how significant their stories are to naval history.

By the early 1900s John Holland had still not given up hope of being the first pioneer of submarines to push his ideas to an actual navy. His sixth design was now introduced to a group of people who were completely different than before – the members of his new Holland Torpedo Boat Company (later to be named the Electric Boat Company). This had been set up following his investments into the American storage battery industry and this had allowed him to work on his projects without any of the previous problems and he soon had the finished boat ready for testing, which was aptly christened USS *Holland*. At long last the American government finally put their previous reservations aside and bought Holland's submarine for $165,000 in the April of 1900. Not only did the Americans sit up and take note of Mr Holland and his designs, the Royal Navy did too.

The fact that Holland was an Irishman and he had already decided to build the *Fenian Ram* as well as selling his concept to the US Navy it would have perhaps got the senior officers of the Royal Navy to sit up and take note when officially these countries were all supposed to be friends and allies; it has to be asked whether it was the threat of these friendly rivals becoming more advanced that caused the minds to change within the Admiralty. This Holland submarine was paid for by the Irish Fenian Society after all, so this could have set the tongues wagging within the planning meetings that meant after all the talk they simply had to adopt some kind of submarine idea before it was too late; if they did not then there was a serious risk of the Royal Navy being left behind and that would have been disastrous at worst, embarrassing at best.

But at last Holland was taken seriously by the Royal Navy and soon the secret project that was to be *Holland I* was born. Four years after the Americans had launched their own version, the British military finally had a rival to show off. Fitted with an internal combustion engine with electric motor and battery, the *Holland I* was launched in 1901 and became the Royal Navy's first submarine to fly the White Ensign, almost four decades after the *USS Housatonic* had been sunk in the first undersea attack. Due to this being the Navy's first submarine, it was very much still an experimental craft, the prototype for the future of the submarine service and one that would be used to iron out any future issues during the testing and trials periods that would now be going ahead, despite much the same design having already been tested and used operationally by the US Navy. Already John Holland had been commissioned to develop another four identical boats, of course they were to be named *Holland II* to *Holland V*. It seemed ironic that the whole reason the submarine came about in this period was due to the anti-British feeling between many of the foreign organisations, the Irish and American submarines being built as a potential weapon against their unofficial enemy, although Holland later fell out with the Irish backers and sold his American design to the British anyway, so it seemed that his loyalty only went as far as his wallet would allow.

The British Holland boats were never really used in anything other than tests and trials, other than perhaps a bit of surveillance here and there; *Holland I* was eventually towed off to be scrapped in 1913 after only twelve years of service and sank on the way to the scrapyard during bad weather off the Cornish coast. Her wreck was located in 1981 and she was salvaged a year later. After many years of restoration she is now open to the public at the Royal Navy Submarine Museum in Gosport, Hampshire. *Holland V* suffered a similar fate off Sussex, although her wreck is still on the seabed and is designated a historic site under the Protection of Wrecks Act 1973.

After many hundreds of years of designs, tests, trials and inevitably disaster, today we have fully functional submarines that allow crew to live and work on board comfortably for several months at a time. The weapons they carry include torpedoes, missiles and in some cases the country's nuclear deterrent. The Royal Navy today has one nuclear powered boat loaded with nuclear weapons at sea at some point 24 hours a day, 365 days a year and has done so since the 1960s. But after the introduction of the Hollands, the race to develop the submarine as a fighting unit went into overdrive and at the start of the First World War, only 13 years since the commissioning of the *Holland I*, submarines were now starting to attack ships as part of the unrestricted underwater warfare. Another milestone in submarine history came when HMS *Pathfinder* became the second ship ever to be sunk by a submarine in action in 1914 after being torpedoed in the North Sea by the German submarine *U-21*. By the start of the Second World War in 1939 the submarine was a common element of the naval fleets of the world; Germany built over 1000 in that six year period, yet lost over 750 of them in just those years alone. Again just 15 years following the end of that war, the nuclear boats were showing the world that they were simply more than just a torpedo carrying stuffy contraption. Between *Holland I* and the launch of HMS *Dreadnought*, just 59 years had passed for them to go from experimental to nuclear powered. Today the most advanced and sophisticated underwater technology now exists to keep fleets and even entire countries safe from attack; collaboration between Australia, the UK and the USA has led to the new AUKUS submarine project, announced in 2023, for the next generation of boats.

As the latest technology sees the launch of the last of the Astute class, and the development of the new Dreadnoughts in the pipeline, we can look back on history and thank those pioneers who came before that have led us to this moment. The Howards, Hollands, Allans and Garretts who put their all in to make something so new and ridiculous that it just might work. And as we can see, it eventually did.

We can only guess over 100 years on what sort of a personality Howard and Allan had, what they were like to talk to, how they conducted themselves in public, what they did in private and how many hours they would spend poring over their work until it was perfected to their standard. There will always be questions over their work and one can only imagine what it must have been like to be in the company of these two gentlemen while they were demonstrating their submarine inventions. Did they know each other? Were each of them aware of the work the other was doing and would they consider themselves rivals? It would be hard to be in such a niche line of work and not know what was going on in that area of expertise. While we can trace the life of James Ellis Howard as well as that of Charles Seymour Allan to a point, we can appreciate what they did in an era that made it difficult to be recognised for what was being done, especially when it came to something so new and untested.

Today we can look back in history and thank them for what they did, for it was people like this that made the navies of the world what they are today.

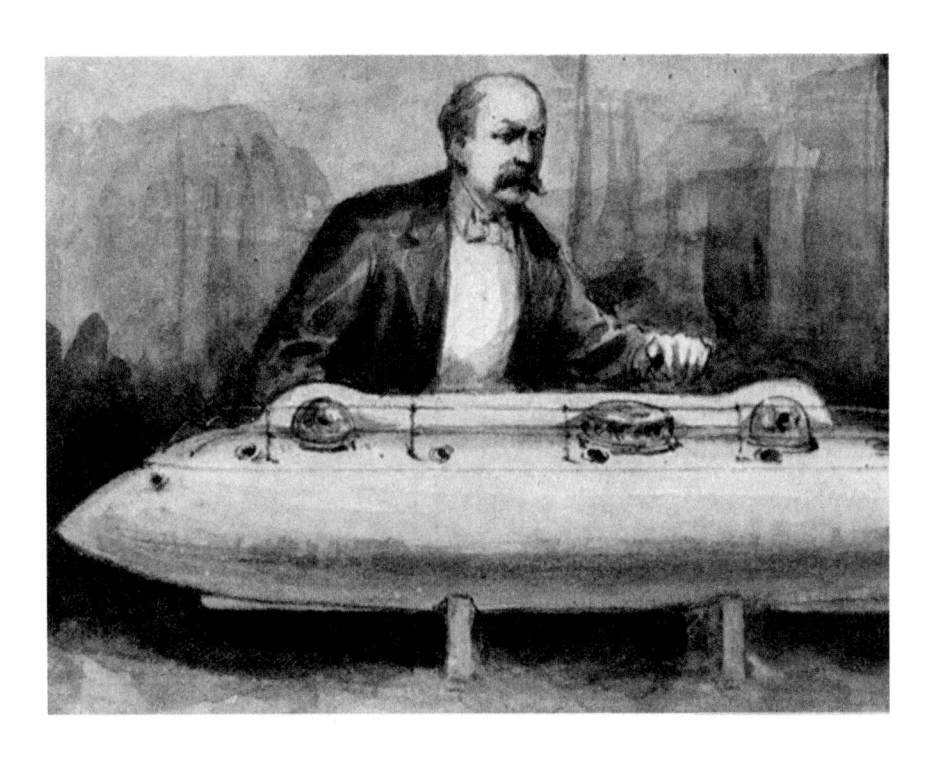

Charles Seymour Allan (c) Illustrated London News Mary Evans Picture Library

Acknowledgements

Daphne Purdon

Dianne Allan

Deb Arnett

Dr Matthew Heaslip, University of Portsmouth

Tasmania Libraries

Mary Evans Picture Library

Sources

This book would not have been possible without the use of the following historic newspapers:

The Argus

The Age (Melbourne)

Bendigo Advertiser

Cobram Courier

Daily Express

Daily Mail

Dundee Courier

Evening Telegraph (Australia)

Illustrated London News

Oamaru Mail

Punch (Melbourne)

Rhyl Record and Advertiser

Sheffield Independent

Sydney Evening News

Zeehan and Dundas Herald (Tasmania)

Author Biography

Richard M. Jones is an author and historian who has made it his life's work to research and highlight forgotten history. Projects have included 16 memorial plaques to date, TV appearances and numerous magazine articles, The Forgotten Submarine Pioneers is his 23rd book. Richard lives with his wife, children and cats and spends most of his time travelling between Hampshire and Yorkshire.

Author Contact

Richard M. Jones

Website:

https://shipwreckdata.wixsite.com/richard-m-jones

Blog: https://disasters-shipwrecks.blogspot.com

Email: shipwreckdata@yahoo.co.uk

Facebook: Richard M Jones

Instagram: @wreckmasterjay

YouTube: @RichardMJones-js7fz

Printed in Great Britain
by Amazon